INTERESTING FACTS FOR INTELLIGENT MEN

INTRODUCTION

Welcome to 'Interesting Facts for Intelligent Men,' a captivating journey delving into the mysteries of our world and beyond. Each page from Planet to Technology unveils a different facet, inviting you to explore the marvels of the Brain, Culture, Space, Literature, and more. This compendium celebrates knowledge, offering portals to new frontiers for curious minds, a mosaic of disciplines inspiring awe for the richness of our universe

About the Author

"I ALWAYS LOVE DISCOVERING NEW FACTS"

CONTENT

PLANET

The Earth formed some 9 billion years after the universe did, making it roughly 4.5 billion years old.

The two main gases that make up Earth's atmosphere are nitrogen (78%) and oxygen (21%), with traces of argon, carbon dioxide, and water vapor.

At roughly 36,000 feet (11,000 meters), the Mariana Trench in the Pacific Ocean is the lowest spot on Earth that is currently known to exist.

There is some variation in the Earth's rotation. The moon and sun's tidal forces are causing it to gradually slow down at an erratic rate.

Just 3% of the water covering more than 70% of Earth's surface is freshwater; the majority is found in glaciers or underground and is not accessible.

The primary elements of the Earth's core are iron and nickel.

The magnetic field of Earth is produced by the separation of the planet's core, which is primarily made of iron and nickel, into a liquid outer core and a solid inner core.

About 100 lightning bolts strike Earth per second, which helps to maintain the electrical equilibrium of the planet and create the ozone layer.

The shape of the Earth's orbit around the sun is elliptical rather than perfect. Our distance from the sun varies as a result throughout the year.

Earth is protected from dangerous solar radiation by the atmosphere. The majority of the sun's UV radiation is absorbed by the ozone layer, which is located in the stratosphere and shields life on Earth from it.

Because of its revolution, the Earth is not a perfect sphere but rather an oblate spheroid that bulges at the equator and is slightly flattened at the poles.

GALAXY

1. Worlds are tremendous frameworks comprising of stars, gas, residue, and dull matter kept intact by gravity. They come in different shapes, including twisting, curved, and sporadic.

2. The Smooth Way, our home world, is a banished twisting system containing between 100 billion to 400 billion stars.

3. Systems are not stale; they're in steady movement. The Smooth Way, for example, is traveling through space at around 1.3 million miles each hour (2.1 million kilometers each hour).

4. The biggest cosmic systems known as "monster ellipticals" can contain trillions of stars and length huge number of light-years across.

5. Systems frequently impact and converge more than billions of years. At the point when systems impact, their stars regularly don't crash because of the huge distances between them, yet their gravitational cooperations can modify their shapes.

6. A few systems, known as dynamic worlds, have supermassive dark openings at their focuses that effectively consume encompassing matter, radiating colossal measures of energy.

7. The discernible universe contains billions of systems. Gauges propose there could be upwards of 2 trillion cosmic systems in the perceptible universe alone.

8. Systems can change fundamentally in their ages. Probably the earliest systems shaped only two or three hundred million years after the Huge explosion.

9. The most far off systems saw by telescopes are billions of light-years away, meaning the light we see from them started its excursion towards us billions of years prior.

10. The Andromeda Cosmic system, situated around 2.537 million light-years away, is on an impact course with the Smooth Way. In around 4 billion years, these two cosmic systems will converge to shape another world.

INVENTIONS

The wheel, considered perhaps of humankind's most urgent development, traces all the way back to around 3500 BC and altered transportation and hardware.

The print machine, imagined by Johannes Gutenberg in the fifteenth hundred years, empowered large scale manufacturing of books and altogether affected the spread of information and data.

The light, frequently connected with Thomas Edison, was really a cooperative development. Edison's group fostered the principal monetarily pragmatic glowing light.

A few innovations have steered history, for example, the steam motor, which powered the Modern Upset and changed assembling and transportation.

The web, created through commitments from different researchers and specialists, changed correspondence, data sharing, and worldwide availability.

Creations like the phone, protected by Alexander Graham Ringer in 1876, reshaped significant distance correspondence and established the groundwork for current broadcast communications.

A few developments were at first met with doubt, similar to the plane. The Wright siblings' effective trip in 1903 reformed travel and transportation.

Advancement frequently expands upon past creations. For example, the improvement of the cell phone incorporated different innovations like processing, correspondence, and cameras into a solitary gadget.

Innovators frequently face difficulties, from financing and suspicion to refining models. Constancy and assurance are fundamental in the development cycle.

CLIMATE

The World's environment has continually changed over its time because of regular elements like volcanic emissions, sun based radiation, and changes in the planet's circle and hub slant.

Human exercises, particularly the consuming of petroleum products, deforestation, and modern cycles, have fundamentally sped up an Earth-wide temperature boost and environmental change somewhat recently.

Environmental change influences different parts of the climate, including climbing worldwide temperatures, softening ice covers and glacial masses, more successive outrageous climate occasions like typhoons and heatwaves, and changing precipitation designs.

The World's air is made fundamentally out of nitrogen (78%) and oxygen (21%), with follow measures of different gases like carbon dioxide, methane, and water fume. Expanded degrees of ozone depleting substances add to a worldwide temperature alteration by catching intensity in the environment.

The Intergovernmental Board on Environmental Change (IPCC) is a worldwide group of researchers laid out to evaluate environmental change and its effects. Their reports act as a basic asset for policymakers around the world.

Environment researchers utilize different intermediaries, including ice centers, tree rings, and dregs layers, to remake past environment conditions and grasp verifiable environment varieties.

Environment variation and relief methodologies expect to decrease ozone harming substance emanations, change to sustainable power sources, safeguard weak biological systems, and foster strong foundation to adapt to environment influences.

Sea fermentation, brought about by expanded

carbon dioxide ingestion by the seas, represents a danger to marine life and biological systems, particularly animals with calcium carbonate shells or skeletons.

Environment exiles are people dislodged from their homes because of environment related factors like ocean level ascent, outrageous climate occasions, or desertification, adding to relocation and cultural difficulties.

BRAIN

The mind weighs around 3 pounds (1.4 kilograms) and contains around 86 billion neurons, the cells answerable for sending data.

Regardless of its somewhat little size, the mind consumes a lot of energy, utilizing around 20% of the body's oxygen and calories.

Brain adaptability permits the mind to revamp itself by framing new brain associations over the course of life, empowering learning, variation, and recuperation from wounds.

The cerebrum produces around 70,000 contemplations on normal each day, showing the unbelievable handling limit of the human brain.

Data goes in the mind at velocities of up to 268 miles each hour (431 kilometers each hour) along brain connections, empowering quick correspondence between various areas.

Dreams happen during the REM (Fast Eye Development) phase of rest, when mind action is high and looks like attentiveness. Dreams assume a part in memory union and close to home handling.

The left half of the globe of the mind is commonly connected with intelligent reasoning, scientific abilities, and language, while the right side of the equator is connected to imagination, instinct, and spatial mindfulness.

The cerebrum's prefrontal cortex, answerable for direction and complex reasoning, keeps creating until the mid-20s, influencing judgment and drive control in teenagers.

Constant pressure can truly change the cerebrum's design and capability, affecting memory, navigation, and profound guideline.

The cerebrum's default mode organization (DMN) enacts when the brain is very still, adding to fantasizing, self-reflection, and imagination, assuming a vital part in inward perspectives.

CULTURE

Culture envelops the convictions, customs, customs, dialects, expressions, and social ways of behaving shared by a specific gathering or society.

Culture isn't stale; it develops and changes over the long run because of impacts from verifiable occasions, globalization, relocation, and innovative progressions.

Social variety alludes to the range of societies coinciding inside a general public or internationally, encouraging shared regard, understanding, and appreciation for various practices and points of view.

Culture impacts individual personalities and ways of behaving, forming how individuals see the world and collaborate with others inside their general public.

Social standards, or acknowledged ways of behaving inside a general public, shift broadl.

across various societies. What's viewed as respectful or OK in one culture might vary fundamentally from another.

Dialects are a necessary piece of culture, with large number of dialects spoken around the world, each conveying its set of experiences and exceptional approaches to offering viewpoints and thoughts.

Social legacy incorporates unmistakable and elusive perspectives like landmarks, curios, dialects, and information went down through ages, adding to a general public's personality.

Globalization has prompted the mixing of societies, bringing about social hybridization, where components from various societies combine, making new customs and practices.

Social trades, for example, craftsmanship displays, and global coordinated efforts, assume a fundamental part in advancing multifaceted comprehension and encouraging appreciation for different social articulations.

LANGUAGE

There are around 7,000 dialects spoken around the world, yet the greater part of the total populace talks only 23 of them as their local language.

Dialects continually develop. New words are added to word references every year, while certain dialects become wiped out, with one language vanishing roughly at regular intervals.

Language shapes the manner in which we think and see the world. A few dialects have explicit words or expressions that don't have direct interpretations into different dialects, reflecting novel social ideas.

The typical individual is assessed to be aware around 20,000 to 35,000 words, yet jargon size fluctuates fundamentally among people in light of elements like schooling, openness, and interests.

Communications through signing are finished and complex dialects with their punctuation and sentence structure. They're not immediate interpretations of communicated in dialects but rather have their own etymological designs.

Dialects are arranged into language families in light of their likenesses. For instance, English has a place with the Germanic language family, which is important for the bigger Indo-European language family.

The most seasoned realized composed language is Sumerian cuneiform, tracing all the way back to around 3200 BC in Mesopotamia (advanced Iraq). Be that as it may, communicated in language originates before composed language by a huge number of years.

Language isn't solely human. Creatures like dolphins, whales, birds, and primates have their types of correspondence, however they contrast essentially from human dialects.

The investigation of how dialects change after some time is known as authentic semantics. It investigates how dialects veer from a typical precursor and tracks their development through hundreds of years.

Language securing in kids happens quickly during the basic time of advancement. Kids presented to different dialects quite early in life can undoubtedly become bilingual or multilingual without disarray, displaying the cerebrum's versatility to language.

HISTORY

The earliest written history traces all the way back to around 3200 BC with the creation of writing in old Mesopotamia, denoting the start of set up accounts.

The Library of Alexandria in old Egypt was one of the biggest and most critical libraries in the old world, containing a huge assortment of original copies and information.

The Dark Passing, quite possibly of the deadliest pandemic in mankind's set of experiences, is assessed to have killed between 75 to 200 million individuals in Europe in the fourteenth 100 years, prompting huge cultural changes.

The Modern Upset, which started in the late eighteenth hundred years in England, changed social orders from agrarian-based economies to industrialized countries, altering the manner in which individuals lived and worked.

The primary fruitful fueled plane flight was

accomplished by the Wright siblings, Orville and Wilbur Wright, on December 17, 1903, at Kitty Falcon, North Carolina, denoting a fantastic crossroads in flying history.

The old city of Petra in Jordan was laid out as soon as 312 BC and was a significant exchanging center point famous for its stone cut design and water channel framework.

The idea of zero as a placeholder in science began in old India around the fifth century Promotion, reforming numerical documentation and estimation.

The Virus War, a time of international strain between the Soviet Association and the US, described by political, military, and philosophical competition, formed worldwide governmental issues from the last part of the 1940s to the mid 1990s.

SPACE

Space isn't totally unfilled; it's loaded up with different types of energy, including electromagnetic radiation like light, X-beams, and inestimable beams.

The Global Space Station (ISS) circles Earth at a typical height of around 250 miles (400 kilometers) and goes at a speed of roughly 17,500 miles each hour (28,000 kilometers each hour).

Stars, cosmic systems, planets, space rocks, comets, and interstellar gas make up a negligible portion of the tremendousness of room.

The biggest known structure known to man is the inestimable web, made out of system bunches and tremendous fibers spreading over a huge number of light-years.

Gravity is the power that administers the developments of heavenly bodies in space, keeping planets in circle around stars and systems bound together.

Stargazers gauge that there could be billions of possibly livable planets in our universe alone, proposing the chance of life past Earth.

Space investigation has prompted various mechanical progressions, including satellite correspondence, GPS frameworks, and clinical developments, helping life on The planet.

ROBOTICS

Mechanical technology consolidates designing, software engineering, and different disciplines to configuration, build, work, and use robots in different applications.

The principal modern robot, Unimate, was presented in 1961 and utilized for errands like stacking hot bits of metal in a pass on projecting plant.

The field of mechanical technology includes different branches, including modern mechanical technology, clinical mechanical technology, social mechanical technology, and bio-roused advanced mechanics, among others.

As innovation progresses, robots are turning out to be more flexible and fit for performing complex errands, from surgeries and investigating distant conditions to aiding debacle aid projects.

Moral contemplations are vital in advanced mechanics, provoking conversations about the effect of man-made intelligence and computerization on business, security, and direction.

Cooperative robots, or cobots, are intended to work close by people, improving efficiency and wellbeing in businesses like assembling and medical care.

The eventual fate of advanced mechanics remembers improvements for delicate advanced mechanics, where robots are made of adaptable and versatile materials, impersonating regular organic entities' developments.

Mechanical technology keeps on developing with developments in simulated intelligence, AI, and tangible advances, prompting headways in independent frameworks fit for learning and adjusting to new conditions and tasks.b

MATHEMATICS

The number zero was freely created by various societies, with the Indian mathematician Brahmagupta being quick to regard it as a number and in addition to a placeholder.

Fibonacci numbers, a succession where each number is the amount of the two going before ones (0, 1, 1, 2, 3, 5, 8...), are tracked down in nature, as in the game plan of petals in blossoms or the expanding of trees.

The idea of limitlessness in science comes in various sizes; there's countable endlessness (like the whole numbers) and uncountable boundlessness (like genuine numbers), prompting amazing thoughts about the tremendousness of numbers.

The disclosure of analytics by Isaac Newton and Gottfried Wilhelm Leibniz reformed arithmetic and laid the preparation for figuring out change and movement.

The Pythagorean hypothesis, which connects with the sides of a right-calculated triangle ($a^2 + b^2 = c^2$), is one of the most seasoned and most central equations in science.

Gödel's Deficiency Hypotheses, formed by mathematician Kurt Gödel, demonstrated that there are valid numerical articulations that can't be demonstrated inside specific proper frameworks, testing total numerical frameworks.

Indivisible numbers, which are distinct simply by 1 and themselves, have entranced mathematicians for a really long time. They are the structure blocks of every single regular number.

Likelihood hypothesis, created by mathematicians like Pierre-Simon Laplace and Blaise Pascal, gives a system to understanding vulnerability and haphazardness, pertinent in different fields like financial matters, physical science, and gaming.

Fractals, mathematical shapes with unpredictable and rehashing designs, are an entrancing area of science. They're tracked down in nature and are created by basic numerical guidelines, yet show intricacy at different scales.

CHEMISTRY

Science is the investigation of issue, its properties, arrangement, and the progressions it goes through. It's frequently called the "focal science" since it interfaces physical science, science, and other innate sciences.

The occasional table, a foundation of science, sorts out components in light of their nuclear number, electron design, and repeating substance properties.

Substance responses include the breaking and framing of compound connections between molecules. Bonds can be covalent, where molecules share electrons, or ionic, where electrons are moved.

Carbon, known as the "building block of life," shapes the premise of natural science, the investigation of mixtures containing carbon-hydrogen bonds.

Substance balance is a state in a reversible response where the forward and switch responses happen at a similar rate, keeping a consistent proportion of reactants and items.

Impetuses are substances that accelerate synthetic responses without being consumed simultaneously. They bring down the enactment energy required for responses to happen.

The pH scale estimates the corrosiveness or basicity of a substance. It goes from 0 to 14, with 7 being impartial. Substances under 7 are acidic, and those over 7 are fundamental.

Components heavier than uranium (nuclear number 92) are normally engineered and are made in research centers through processes like atomic combination or molecule barrage.

Electrochemistry concentrates on the connection between synthetic responses and power, like in batteries or electrolysis, where electrical energy drives compound changes.

Science assumes a urgent part in daily existence, from the food we eat, the prescriptions we take, and the materials utilized in innovation and development, to the natural cycles molding our reality.

NEUROSCIENCE

The mind comprises of around 86 billion neurons, each equipped for associating with large number of different neurons, framing a complicated organization that controls contemplations, feelings, and physical processes.

The cerebrum utilizes around 20% of the body's oxygen and energy regardless of representing just around 2% of its complete weight.

The sensory system has two primary divisions: the focal sensory system (cerebrum and spinal rope) and the fringe sensory system (nerves outside the mind and spinal string).

Neurotransmitters are the intersections between neurons where data is communicated. There are assessed to be trillions of neural connections in the human cerebrum.

Synapses are substance couriers that communicate signals across neural connections. Normal synapses incorporate dopamine, serotonin, and acetylcholine, each assuming explicit parts in mind capability.

The prefrontal cortex, liable for direction and complex mental way of behaving, is among the last cerebrum districts to develop, proceeding with improvement into an individual's mid-20s.

Mind imaging advancements like X-ray (Attractive Reverberation Imaging) and fMRI (useful X-ray) permit researchers to picture cerebrum designs and action, giving experiences into cerebrum capability.

The investigation of mirror neurons recommends that specific neurons in the cerebrum are enacted both when an individual plays out an activity and when they notice another person playing out a similar activity, possibly adding to sympathy and social way of behaving.

Problems like Alzheimer's, Parkinson's, schizophrenia, and mental imbalance are connected to disturbances in mind design, capability, or synapse movement, featuring the intricacy of neurological circumstances.

LITERATURE

The most established realized composed story is the "Epic of Gilgamesh," tracing all the way back to old Mesopotamia. It's viewed as one of the earliest enduring works of writing.

Shakespeare, frequently viewed as perhaps of the best writer and artists ever, authored various words adages actually utilized in the English language today.

The novel, as a particular scholarly structure, arose in the eighteenth 100 years with works as defoe Daniel's "Robinson Crusoe" and Samuel Richardson's "Pamela."

Numerous scholarly works have been restricted or controlled since the beginning of time because of multiple factors, including political, strict, or social complaints.

A portion of the world's most famous scholarly works, similar to "The Odyssey" and "The Iliad" by

Homer, were initially oral customs prior to being translated into composed structure.

Certain books, for example, George Orwell's "1984" and Aldous Huxley's "Exciting modern lifestyle," have been astoundingly perceptive in anticipating cultural and mechanical headways.

Writing can act as a useful asset for social change, supporting for equity, uniformity, and common freedoms. Works like Harriet Beecher Stowe's "Uncle Tom's Lodge" assumed a critical part in the abolitionist development.

The Nobel Prize in Writing, laid out in 1901, perceives exceptional commitments to writing. Past laureates incorporate famous creators like Gabriel García Márquez, Toni Morrison, and Ernest Hemingway.

Scholarly types have developed after some time, incorporating a great many styles and structures, including fiction, verse, show, true to life, dream, sci-fi, and the sky is the limit from there, reflecting different human encounters and minds.

SPORTS

Sports have existed for centuries, profoundly imbued in human culture across different developments.

Soccer (football) gloats the biggest worldwide fanbase and player cooperation.

The starting points of the long distance race date back to a Greek trooper, Pheidippides, who went around 26 miles (42 kilometers) from Long distance race to Athens to convey fresh insight about a tactical triumph.

B-ball was created in 1891 by James Naismith, utilizing a soccer ball and peach containers.

The cutting edge Summer Olympics, which started in 1896, highlighted nine games; today, they incorporate more than 30 games.

Cricket, among the most seasoned group activities, has attaches tracing all the way back to the sixteenth hundred years and is generally well known in numerous nations.

Certain games, similar to aerobatic, figure skating, and plunging, frequently see maximized operations in competitors during their late youngsters or mid twenties because of the actual requests and accuracy required.

The Olympic Games, following back to antiquated Greece, were once again introduced in the cutting edge time in 1896 to join competitors from around the world in amicable contest.

ARCHITECTURE

Engineering is something other than planning structures; it's a mix of craftsmanship, science, and usefulness, forming the actual climate where individuals reside, work, and play.

The Incomparable Pyramid of Giza in Egypt, worked around 2560 BCE, was the tallest man-made structure for north of 3,800 years and is the main enduring Seven Marvels of the Antiquated World.

Gothic design, pervasive in archaic Europe, is described by pointed curves, ribbed vaults, and flying braces, considering taller and more unpredictable church structures.

Blunt Lloyd Wright, a compelling American planner, presented the idea of natural engineering, accentuating agreement among structures and their regular environmental factors, seen in Fallingwater and the Guggenheim Exhibition hall.

The Eiffel Pinnacle in Paris, built for the 1889 World's Fair, was at first censured by a larger number of people however has turned into a famous image of France and a wonder of designing.

The Colosseum in Rome, finished in Promotion 80, was an amphitheater fit for seating 50,000 to 80,000 observers and facilitated warrior challenges, creature chases, and other public displays.

Brutalist design, famous during the twentieth 100 years, is described by crude substantial surfaces, mathematical shapes, and an emphasis on usefulness, exemplified by structures like the Barbican Place in London.

Antiquated human advancements, similar to the Incas and Mayans, made amazing compositional accomplishments without present day innovation, like Machu Picchu and the sanctuaries of Chichen Itza.

The utilization of supportable and eco-accommodating plan standards in contemporary design means to diminish natural effect, using sustainable assets and energy-productive materials.

Contemporary engineers frequently consolidate inventive advances like parametric plan and 3D printing to make multifaceted, productive, and interesting designs, pushing the limits of conventional compositional standards.

ASTRONOMY

Cosmology is the investigation of divine items and peculiarities outside Earth's climate, including stars, planets, worlds, and vast occasions.

The Hubble Space Telescope, sent off in 1990, has given cosmologists amazing pictures and information about far off systems, stars, and nebulae, reforming how we might interpret the universe.

Cosmologists utilize different devices and methods, like telescopes, spectroscopy, and space tests, to notice, measure, and dissect divine items and peculiarities.

Cosmologists have recognized a huge number of exoplanets (planets outside our planetary group) utilizing strategies like the travel technique and spiral speed estimations. A portion of these exoplanets could have conditions reasonable forever.

Stargazing has various branches, including astronomy, cosmology, planetary science, and observational stargazing, each zeroing in on unambiguous parts of the universe.

Nebulae, tremendous billows of gas and residue in space, are heavenly nurseries where new stars are conceived. They come in different shapes and tones, frequently making shocking divine vistas.

The investigation of light discharged or consumed by divine items, known as spectroscopy, permits cosmologists to decide the synthetic structure, temperature, and different properties of stars and universes.

The electromagnetic range, which incorporates radio waves, microwaves, infrared, noticeable light, bright, X-beams, and gamma beams, is instrumental in grasping divine items, as various frequencies give one of a kind data about them.

GEOGRAPHY

Topography envelops the investigation of Earth's scenes, surroundings, and the connections among individuals and their environmental factors.

The world's biggest desert isn't the Sahara; it's Antarctica. In spite of being shrouded in ice, it gets next to no precipitation, qualifying it as a desert.

The Pacific Sea isn't simply the biggest sea yet in addition covers more region than every one of the World's bodies of land consolidated.

The Incomparable Obstruction Reef in Australia is the biggest coral reef framework all around the world and might actually be seen from space.

Mount Everest is the tallest mountain above ocean level, yet the tallest mountain estimated from base to highest point is Mauna Kea in Hawaii, as its vast majority is submerged.

The Amazon Rainforest, frequently called the "lungs of the Earth," produces around 20% of the world's oxygen and houses an amazingly different environment.

The Sahara Desert grows at a pace of around 0.8 kilometers (0.5 miles) toward the south every year because of desertification brought about by environmental change and human exercises.

Topography impacts societies and social orders. The appropriation of assets, environment designs, and topographical elements shape how civilizations create and flourish across the globe.

GEOLOGY

Geography is the investigation of the World's construction, cycles, materials, and history, including its stones, minerals, fossils, and landforms.

The World's hull is separated into structural plates that float on the semi-liquid asthenosphere. The development of these plates causes tremors, volcanic emissions, and the arrangement of mountain ranges.

Rocks are grouped into three principal types: volcanic, sedimentary, and transformative. Volcanic rocks structure from liquid magma, sedimentary rocks from compacted and established dregs, and transformative rocks from existing rocks changed by intensity and tension.

Fossils give significant insights about previous existence on The planet. They are protected remaining parts or hints of living beings tracked down in sedimentary rocks, offering bits of knowledge into old conditions and transformative history.

The World's center, made for the most part out of iron and nickel, produces an attractive field that stretches out into space, safeguarding the planet from hurtful sun based radiation.

Plate tectonics drive the constant course of rock development, disintegration, and reusing known as the stone cycle, forming the World's surface north of millions of years.

Geologists utilize different dating methods, including radiometric dating, to decide the time of rocks and fossils by investigating the rot of radioactive components inside them.

Mountains structure through various land cycles like volcanic movement, structural plate crashes, or the inspire of crustal blocks because of powers inside the Earth.

Minerals are normally happening inorganic substances with explicit compound sytheses and gem structures. They are the structure blocks of rocks and have different purposes in industry, innovation, and day to day existence.

Geography assumes an essential part in understanding regular dangers like avalanches, waves, and volcanic emissions, empowering researchers to survey gambles and foster procedures to moderate their effect on networks.

TECHNOLOGY

Innovation has dramatically progressed throughout the long term. The cell phone in your pocket is more impressive than the PCs that directed the Apollo missions to the moon.

The expression "innovation" starts from the Greek words "techne" (workmanship, expertise) and "logia" (investigation of), basically meaning the investigation of craftsmanship, ability, or specialty.

The primary PC mouse was developed during the 1960s by Douglas Engelbart. It was made of wood and had just a single button.

The Internet was concocted by Tim Berners-Lee in 1989. It opened up in 1991, changing worldwide correspondence and data sharing.

The idea of 3D printing traces all the way back to the 1980s, yet it acquired far reaching consideration and application in different ventures, including medical services, assembling, and aviation, in the mid 21st hundred years.

Quantum processing, which outfits the standards of quantum mechanics, holds the possibility to take care of intricate issues dramatically quicker than old style PCs by utilizing quantum bits or qubits.

Computerized reasoning (man-made intelligence) is a quickly developing field that includes machines emulating human mental capabilities. Man-made intelligence advancements power menial helpers, self-driving vehicles, and prescient investigation.

Innovation has essentially affected medical services through developments like telemedicine, automated medical procedures, customized medication, and wearable wellbeing checking gadgets.

INNOVATION

Advancement isn't just about developments; it's likewise about imaginatively applying existing thoughts or innovations in better approaches to tackle issues or address issues.

Disappointment is many times a significant piece of the development cycle. Numerous effective advancements arose out of various bombed endeavors and cycles.

Cooperative advancement, including different points of view and skill, frequently prompts really weighty and compelling arrangements.

Developments can emerge from different sources, including logical examination, innovative headways, social changes, and, surprisingly, unforeseen blends of existing thoughts.

Persistent advancement is fundamental for remaining serious in quickly developing enterprises, driving monetary development and cultural advancement.

Advancements can have unseen side-effects, both positive and negative, molding ventures, economies, and social orders in unexpected ways.

Steady advancement includes making little enhancements or changes to existing items or cycles, while problematic development in a general sense modifies existing business sectors or enterprises.

Development flourishes in conditions that empower trial and error, risk-taking, and a culture that values imagination and gaining from botches.

Licensed innovation freedoms, like licenses, copyrights, and brand names, assume a part in securing and boosting development by giving select privileges to makers.

Development isn't restricted to innovation; it reaches out to regions, for example, plans of action, authoritative designs, social frameworks, and inventive expressions, impacting different parts of human existence.

HEALTH

Normal actual work can diminish the gamble of persistent sicknesses like coronary illness, diabetes, and certain diseases. Indeed, even modest quantities of activity can have a major effect.

The stomach microbiome, containing trillions of microscopic organisms and microorganisms in our stomach related framework, assumes an essential part in generally speaking wellbeing, influencing processing, resistance, and, surprisingly, psychological well-being.

Sufficient rest is fundamental for good wellbeing. It's during rest that the body fixes and revives itself, supporting mental capability, temperament, and in general prosperity.

Ongoing pressure can adversely affect wellbeing by adding to conditions like hypertension, debilitated insusceptible framework, and expanded chance of psychological well-being issues.

Adjusted nourishment is key for ideal wellbeing. An eating regimen wealthy in organic products, vegetables, entire grains, lean proteins, and solid fats gives fundamental supplements to the body.

Emotional well-being is just about as significant as actual wellbeing. Dealing with mental prosperity through rehearses like care, reflection, or treatment can significantly affect by and large wellbeing.

The resistant framework protects the body against contaminations and sicknesses. Keeping a solid way of life with legitimate sustenance, exercise, and rest upholds major areas of strength for a framework.

Customary wellbeing screenings and check-ups are pivotal for early identification and counteraction of illnesses. They can frequently get potential medical problems before they become serious.

Social associations and connections essentially affect wellbeing. Solid social help can work on psychological well-being and even upgrade actual wellbeing results.

PANDEMICS

Pandemics are boundless flare-ups of irresistible infections that influence a huge geological region, frequently crossing various nations or mainlands.

The Dark Demise, perhaps of history's deadliest pandemic, happened in the fourteenth 100 years and cleared out an expected 75-200 million individuals in Eurasia, radically modifying society.

The 1918 Flu Pandemic, known as the Spanish influenza, tainted around 33% of the total populace and caused an expected 50 million passings universally.

Pandemics can start from different sources, including infections, microscopic organisms, and different microbes. The transmission can happen through the air, water, or direct contact.

Coronavirus, brought about by the novel Covid SARS-CoV-2, turned into a pandemic in 2020, prompting boundless sickness, financial disturbance, and worldwide wellbeing challenges.

Over the entire course of time, pandemics have had significant social, monetary, and political effects, reshaping social orders, economies, and medical services frameworks.

The speed and degree of movement today can add to the quick spread of pandemics, as individuals can get across mainlands in practically no time.

Immunizations play had a significant impact in fighting pandemics by giving resistance to populaces against irresistible illnesses.

General wellbeing measures like quarantine, social separating, and cleanliness rehearses are imperative in controlling the spread of pandemics and diminishing their effect.

Pandemics have frequently catalyzed progressions in medication, general wellbeing foundation, and logical examination, prompting better readiness and reactions to future episodes.

NUTRITION

Supplements are separated into macronutrients (carbs, proteins, fats) and micronutrients (nutrients and minerals) fundamental for ideal physical process.

The body requires an equilibrium of macronutrients to work appropriately, with starches giving energy, proteins supporting tissue fix, and fats aiding supplement ingestion and chemical creation.

Micronutrients like nutrients and minerals assume significant parts in different physical processes, from supporting the safe framework to advancing bone wellbeing and helping with digestion.

Fiber, found in plant-based food varieties, is significant for stomach related wellbeing and can decrease the gamble of specific illnesses like coronary illness and diabetes.

Water is a fundamental supplement that makes up a huge part of the body and is pivotal for different physical processes, including managing temperature and helping with assimilation.

Cell reinforcements, present in many products of the soil, assist with combatting oxidative pressure in the body and may lessen the gamble of ongoing illnesses like malignant growth and coronary illness.

The nature of food matters as much as amount. Picking supplement thick food sources like entire grains, lean proteins, and bright products of the soil is key for ideal wellbeing.

Dietary requirements fluctuate among people in light of variables like age, sex, movement level, and generally wellbeing. There's nobody size-fits-all way to deal with sustenance.

Handled food sources frequently contain high measures of added sugars, undesirable fats, and sodium, which can add to different medical problems when devoured in abundance.

The significance of adjusted nourishment stretches out past actual wellbeing. A sound eating routine can decidedly influence mind-set, mental capability, and generally prosperity.

FITNESS

Customary actual work can further develop mind-set by delivering endorphins, which are regular temperament lifters, prompting diminished pressure and tension.

Bulk will in general downfall with age, beginning as soon as in your 30s. Strength preparing can assist with combatting this normal misfortune and keep up with bulk.

Steady work-out is connected to all the more likely rest quality, directing rest designs and work on in general serenity.

Practice helps mind wellbeing by expanding the development of neurochemicals that help discernment and memory, possibly decreasing the gamble of mental degradation as we age.

HIIT (Stop and go aerobic exercise) can give critical wellness benefits in more limited periods contrasted with conventional consistent state exercises, working on cardiovascular wellbeing and calorie consuming.

Adaptability works out, similar to yoga or extending schedules, can upgrade joint versatility, diminish the gamble of injury, and work on in general practical development.

Active work can emphatically influence stomach wellbeing by advancing a different microbiome, possibly lessening the gamble of gastrointestinal issues.

Standard activity has been displayed to increment life expectancy and work on generally personal satisfaction, decreasing the gamble of constant sicknesses like coronary illness, diabetes, and certain tumors.

Rest and recuperation are essential parts of wellness. Muscles need time to fix and develop further after exercises, accentuating the significance of sufficient rest between meetings.

Wellness isn't just about work out; legitimate sustenance assumes a huge part. A fair eating regimen that incorporates proteins, sound fats, carbs, nutrients, and minerals is fundamental for ideal wellness and in general wellbeing.

EDUCATION

Training isn't restricted to formal tutoring; it incorporates long lasting learning, casual encounters, and independent review, adding to individual and expert turn of events.

The idea of instruction has developed over hundreds of years, adjusting to social, innovative, and cultural changes, molding how information is conferred and obtained.

Schooling assumes a urgent part in encouraging decisive reasoning, critical thinking abilities, and imagination, enabling people to adjust to different difficulties in different fields.

Different instructive ways of thinking and strategies exist around the world, going from customary talk based ways to deal with experiential learning, and online schooling.

Admittance to quality schooling stays a worldwide test, with variations in assets, framework, and open doors influencing a great many people, especially in underserved networks.

Innovative progressions have changed training, empowering on the web stages, computerized assets, and intuitive devices that work with advancing past conventional study hall settings.

Instruction isn't exclusively about scholastics; it incorporates social-profound learning, character advancement, and the development of compassion and relational abilities.

Deep rooted learning is progressively fundamental in the quickly advancing position market, provoking the requirement for constant ability improvement and versatility all through one's vocation.

Instruction is an impetus for cultural advancement, cultivating resistance, inclusivity, and understanding among different networks, adding to social attachment.

Quality schooling significantly affects economies, wellbeing, and generally speaking cultural prosperity, filling in as a foundation for individual strengthening and aggregate headway.

MEMORY

Memory is certainly not a solitary element however a perplexing framework including different cycles, including encoding, stockpiling, and recovery of data.

Transient memory considers the impermanent maintenance of data, going on for around 15-30 seconds except if it's moved to long haul memory through practice or affiliation.

Long haul memory is separated into express (cognizant) and certain (oblivious) memory. Express memory incorporates realities and occasions (semantic and roundabout memory), while certain memory includes abilities and propensities.

Recollections aren't fixed; they can be affected by different variables, including feelings, discernments, and resulting encounters, prompting adjustments or recreations.

The mind doesn't store recollections like a video recorder; all things being equal, it remakes them while made, recollections defenseless to mutilation or neglecting.

Memory is significant for character development, as it makes an identity by holding individual encounters and learned information.

Rest assumes a vital part in memory combination. During rest, the mind cycles and moves recently obtained data from present moment to long haul memory.

Certain cerebrum structures, similar to the hippocampus, are crucial for framing new recollections, while the prefrontal cortex helps in coordinating and recovering them.

Neglecting is a characteristic part of memory. It happens because of impedance from new data, rot over the long haul, or recovery disappointment.

Memory can be worked on through different procedures and practices, like mental helpers, ordinary activity, sufficient rest, and mental activities that challenge the cerebrum.

CREATIVITY

Imagination includes consolidating existing thoughts, ideas, or components in better approaches to create novel arrangements or articulations.

It's not restricted to human expression; imagination is fundamental in critical thinking, science, innovation, business, and daily existence.

Studies recommend that cultivating a favorable climate, embracing interest, and considering different points of view can improve and empower imagination.

Innovativeness frequently includes both dissimilar reasoning (creating various thoughts) and concurrent reasoning (assessing and refining those thoughts).

Being available to disappointment and embracing botches as a feature of the inventive strategy can prompt leap forwards and development.

The cerebrum's innovative flow includes different districts cooperating, including the prefrontal cortex for thought age and the hippocampus for memory and affiliation.

Innovativeness can be improved through practices like care, contemplation, openness to new encounters, and various boosts.

Cooperative conditions, where people with various foundations and mastery meet up, frequently flash more savvy fixes.

Inventiveness will in general thrive when there's a harmony among design and opportunity, giving rules while permitting to trial and error.

Imagination is certainly not a decent characteristic; an expertise can be created and worked on through training, investigation, and persistent learning.

SUCCESS

Achievement is definitely not a one-size-fits-all idea. It differs significantly from one individual to another and can envelop accomplishments in different parts of life, including profession, connections, self-improvement, and that's just the beginning.

Determination and versatility are key parts of progress. Numerous fruitful people confronted various disappointments prior to arriving at their objectives.

Achievement frequently includes putting forth clear and feasible objectives, as well as making a guide or plan to arrive at those objectives.

Persistent learning and flexibility are critical for progress in a quickly impacting world. Fruitful people frequently embrace a development mentality and look for new information and abilities.

Systems administration and building connections can altogether add to progress. Associations and joint efforts frequently open ways to open doors.

Using time effectively and prioritization assume an imperative part in making progress. Successfully dealing with one's time and zeroing in on high-need assignments can prompt more noteworthy achievements.

Disappointment is in many cases a venturing stone to progress. Gaining from missteps and involving them as illustrations can make ready for future accomplishments.

The capacity to understand people on a profound level, including mindfulness and the capacity to deal with feelings, is significant for exploring difficulties and encouraging achievement, particularly in relational connections and positions of authority.

Achievement is many times joined by areas of strength for an ethic and devotion. Investing steady energy and remaining focused on one's objectives can yield huge outcomes.

Achievement is emotional and can advance over the long haul. What might be viewed as progress at one phase of life could contrast at another, reflecting changing needs and goals.

POLITICS

Governmental issues is the interaction by which choices are made inside gatherings or social orders, including the conveyance of force, assets, and administration.

"Governmental issues" comes from the Greek word "polis," and that implies city-state. It initially alluded to undertakings of the city-state in antiquated Greece.

Manipulating is a training where political limits are controlled to lean toward one party over another, influencing the result of races.

The political scene frequently includes campaigning, where people or gatherings endeavor to impact political choices, regularly by supporting for explicit approaches or interests.

The idea of a majority rules government, where residents have something to do with dynamic through casting a ballot, traces all the way back to old Athens, thought about the origination of a vote based system.

Global relations incorporate the cooperations between nations, including discretion, arrangements, economic accords, and clashes.

Political missions use different systems, including promoting, rallies, discussions, and web-based entertainment, to convince electors and gain support for applicants or strategies.

Political polarization alludes to the rising split between various ideological groups or philosophies, prompting more outrageous perspectives and less split the difference.

Political frameworks shift around the world, from vote based systems to absolutisms, governments to republics, each with its own design and instruments for administration.

ECONOMICS

Financial matters is the investigation of how people, organizations, and social orders allot assets to fulfill their requirements and needs.

The law of organic market is a central standard in financial matters. It expresses that costs rise when request surpasses supply and fall when supply surpasses request.

GDP (Gross domestic product) gauges the complete worth of labor and products created in a country inside a particular period. It's a critical mark of an economy's wellbeing.

Expansion is the rate at which the general degree of costs for labor and products rises, bringing about a reduction in buying control after some time.

Joblessness rates mirror the level of individuals in the labor force who are effectively looking for business yet haven't gotten a new line of work. It's a pivotal proportion of financial wellbeing.

Economies can be characterized into various sorts, like market economies (where not set in stone by organic market) and order economies (where the public authority controls creation and evaluating).

The investigation of social financial matters investigates how mental variables impact monetary choices, frequently veering off from customary monetary models in view of judiciousness.

Monetary frameworks differ around the world, including free enterprise (confidential responsibility for), communism (aggregate possession), and blended economies (mixing components of both).

Worldwide exchange permits nations to represent considerable authority in creating labor and products they are generally effective at, prompting expanded worldwide proficiency and admittance to different items.

WILDLIFE

Natural life alludes to all non-trained creatures, including warm blooded animals, birds, reptiles, creatures of land and water, bugs, and marine life, living in different biological systems around the world.

Biodiversity inside untamed life is staggeringly assorted. It incorporates an expected 8.7 million species, albeit many are yet to be found and depicted by researchers.

A few animal categories, known as cornerstone species, assume a pivotal part in keeping up with their environments' equilibrium. For example, honey bees are indispensable pollinators, and wolves assist with directing prey populaces.

Natural life faces various dangers, including environment misfortune because of deforestation, urbanization, contamination, environmental change, poaching, and unlawful untamed life exchange.

The preservation of untamed life is fundamental for keeping up with biological equilibrium. Safeguarded regions like public stops and holds mean to save living spaces and protect imperiled species.

Movement is a typical way of behaving among numerous types of untamed life, like birds, fish, and warm blooded creatures, frequently determined via occasional changes or the quest for food and favorable places.

Numerous natural life species show intriguing ways of behaving and transformations, from the perplexing moves of birds during romance to the disguise procedures of different bugs and reptiles.

The endurance of specific species depends on advantageous connections. For example, cleaner fish help bigger marine species by eliminating parasites from their bodies.

Human action has caused the eradication of various natural life species, like the traveler pigeon and the Tasmanian tiger, featuring the effect of human activities on biodiversity.

Preservation endeavors, including living space reclamation, hostage reproducing projects, and local area based protection drives, assume a crucial part in safeguarding and saving untamed life for people in the future.

DISCOVERY

Disclosure frequently emerges from interest and perception, starting investigation and examination concerning obscure or unexplained peculiarities.

Numerous leading edge revelations were coincidental, similar to the disclosure of penicillin by Alexander Fleming in 1928, who saw shape restraining bacterial development on a petri dish.

Logical disclosures frequently go through a thorough course of perception, speculation, trial and error, and examination, adding to how we might interpret the regular world.

Disclosures can happen in assorted fields, from science and innovation to craftsmanship, culture, history, and then some, forming our insight and discernments.

Coordinated effort and sharing of information among researchers and scientists worldwide speed up the speed of disclosure and advancement.

The course of disclosure isn't generally straight; it can include misfortunes, bombed analyzes, and amended speculations prior to arriving at forward leaps.

A few revelations have upset whole ventures, for example, the disclosure of power and ensuing creations like the light by Thomas Edison.

Revelation is a progressing and consistent cycle, prompting new inquiries and areas of investigation, adding to the advancement of human information and understanding.

ENTREPRENEURSHIP

Business venture isn't exclusively about beginning a business; it's likewise about recognizing potential open doors, facing challenges, and executing imaginative plans to make esteem.

Business people frequently embrace disappointment as a growth opportunity. Numerous fruitful business people confronted mishaps prior to accomplishing their objectives.

Business isn't restricted to a particular age bunch. The absolute best business visionaries began their endeavors at different phases of life, from youthful adulthood to later years.

Fruitful business people are much of the time adroit at systems administration and building connections. Associations with coaches, financial backers, and different business visionaries can be pivotal for development.

Innovation has essentially affected business venture. It has brought down section boundaries, permitting more people to begin organizations from anyplace on the planet.

Numerous business people add to social effect through their endeavors. Social business venture centers around resolving cultural issues while keeping up with productivity.

Versatility and readiness are key attributes of fruitful business people. They should explore changing economic situations and turn their techniques as needs be.

Admittance to subsidizing is fundamental for some business visionaries. While some bootstrap their endeavors, others look for speculation through different channels like funding, private backers, or crowdfunding.

Business encourages development and drives monetary development by presenting new items, administrations, and plans of action that meet advancing customer needs.

The enterprising excursion is many times testing yet fulfilling. It requires tirelessness, commitment, and a readiness to consistently learn and develop.

TRAVEL

The world's most established travel service, Thomas Cook, was established in 1841, at first contribution train journeys to decrease liquor utilization among assembly line laborers in Britain.

"Travel" starts from the French word "travailler," and that means to work or work. This derivation mirrors the previous relationship of movement with challenging excursions.

The idea of movement protection traces all the way back to the beginning of the Roman Domain, where internment clubs existed to assist with covering burial service costs for individuals kicking the bucket while away from home.

The popular Silk Street, crossing more than 4,000 miles, associated China with the Mediterranean, encouraging exchange and social trade among East and West.

Going via train is one of the most harmless to the ecosystem ways of voyaging, with essentially lower fossil fuel byproducts contrasted with vehicles or planes.

The movement business contributes fundamentally to the worldwide economy, producing position and pay. In 2019, it was assessed to represent around 10% of worldwide Gross domestic product.

EXPLORATION

Investigation has been a central human undertaking for centuries, tracing all the way back to old developments like the Phoenicians, who were gifted mariners and brokers.

The Time of Investigation, beginning in the fifteenth 100 years, denoted a time of critical sea revelations, including Christopher Columbus' journeys to the Americas.

The primary effective circumnavigation of the Earth was finished by Ferdinand Magellan's undertaking in the sixteenth hundred years, however Magellan himself didn't endure the whole excursion.

Space investigation started during the twentieth 100 years with the send off of Sputnik 1 by the Soviet Association in 1957, denoting the beginning of the space age.

The Apollo 11 mission in 1969 was whenever people first arrived on the moon, with space travelers Neil Armstrong and Buzz Aldrin turning into the first to stroll on its surface.

The investigation of the remote ocean has uncovered shocking biodiversity and biological systems, with just a negligible part of the sea's profundities investigated up until this point.

Polar investigation, strikingly campaigns toward the North and South Poles, has been trying because of outrageous atmospheric conditions and tough territory.

Automated investigation, through tests and wanderers, has extended how we might interpret divine bodies like Mars, Saturn's moons, and comets.

Investigation isn't restricted to actual wildernesses; it additionally incorporates logical exploration, mechanical development, and the mission for information in different fields.

Current investigation includes worldwide joint efforts, for example, the Global Space Station, where nations cooperate to push the limits of human information and capacity.

ADVENTURE

Experience isn't just about outrageous exercises; it's likewise about getting out of your usual range of familiarity to investigate the obscure and encounter self-awareness.

The idea of experience changes enormously among people; what may be an undertaking for one individual could be normal for another.

Experience can come in many structures, from actual pursuits like hiking and skydiving to scholarly difficulties, social investigations, or in any event, attempting new foods.

Taking part in undertakings frequently prompts expanded imagination and critical thinking abilities, as well as improved versatility to new circumstances.

The adrenaline rush experienced during an undertaking triggers the arrival of endorphins, which can support temperament and make a feeling of elation.

Experience travel has become progressively well known, with individuals looking for special encounters like hiking through far off locales, natural life safaris, or investigating antiquated ruins.

Mental examinations recommend that integrating normal dosages of experience into life can work on by and large bliss and fulfillment.

Experience isn't only for the youthful; individuals of any age take part in courageous exercises, demonstrating that it's an outlook in excess of an actual capacity.

The soul of experience frequently implies going ahead with potentially dangerous courses of action, embracing vulnerability, and gaining from the two triumphs and disappointments.

Experience isn't exclusively about the objective; the actual excursion, loaded up with unforeseen exciting bends in the road, frequently holds the most important encounters and examples.

HUMAN BEHAVIOR

Reflect neurons in the mind permit people to identify comprehend others' feelings by reflecting their activities and sentiments, cultivating social associations.

The spectator impact happens when people are more averse to offer assistance in a crisis circumstance when others are available, expecting another person will mediate.

A self-influenced consequence outlines how confidence in a treatment's viability can prompt genuine physiological changes, displaying the brain's strong impact over the body.

People will generally take part in congruity, modifying their ways of behaving or suppositions to line up with bunch standards or cultural assumptions, frequently to keep away from social dismissal or gain acknowledgment.

Securing predisposition alludes to the inclination to depend vigorously on the primary snippet of data experienced (the "anchor") while deciding, regardless of whether it's unessential or erroneous.

The impact of self-fulfilling prophesy features how better standards set upon people can prompt superior execution, proposing that convictions about somebody's capacities can impact their results.

The simple openness impact recommends that individuals will generally foster an inclination for things just in light of the fact that they are know about them, regardless of whether they deliberately review the openness.

The Zeigarnik impact makes sense of how incomplete or hindered errands will generally be recalled more effectively than finished ones, prompting a psychological obsession with deficient undertakings.

HUMER

Humor is a general human encounter found across societies and social orders, filling in as a social device for correspondence and holding.

Chuckling, a typical reaction to go along with, has various medical advantages, including decreasing pressure chemicals and working on invulnerable capability.

Humor can be arranged into different sorts, like droll, wit, parody, observational, and humble humor, interesting to various crowds.

Humor frequently includes disjointedness - the unforeseen or a deviation based on what is standard or expected, making shock and entertainment.

The capacity to appreciate and make humor includes complex mental cycles that draw in different districts of the cerebrum, including regions liable for language, social discernment, and feeling.

Humor can be utilized as a survival strategy in testing circumstances, assisting people with managing pressure, tension, or troublesome feelings.

Various societies have remarkable styles and inclinations in humor, impacted by their qualities, customs, and verifiable setting.

Humor has been utilized in different types of media, including writing, film, TV, and web-based entertainment, to engage, teach, and pass on messages.

Humor isn't generally all around comprehended or appreciated; what one individual finds entertaining probably won't resound with one more because of individual contrasts and social settings.

Fostering a funny bone is a powerful cycle impacted by childhood, encounters, and openness to various types of satire, adding to individual characters and social communications.

Cryptography

Cryptography is the workmanship and study of secure correspondence, utilizing strategies to encode messages such that main approved gatherings can understand them.

The earliest known utilization of cryptography traces all the way back to antiquated developments like Egypt and Mesopotamia, where basic replacement figures were utilized to encode messages.

One of the most popular authentic encryption gadgets is the Conundrum machine, utilized by the Germans during The Second Great War to scramble delicate military correspondences. It was in the long run translated by Alan Turing and his group at Bletchley Park.

The Caesar figure, named after Julius Caesar, is a replacement figure where each letter in the plaintext is moved a specific number of spots down or up the letter set. It's one of the least complex and earliest known encryption strategies.

Cryptography is fundamental in getting present day correspondence and information transmission over the web, including email encryption, secure web perusing (HTTPS), and advanced marks.

Public-key cryptography, imagined during the 1970s, changed secure correspondence. It includes a couple of keys: a public key utilized for encryption and a confidential key for unscrambling.

The High level Encryption Standard (AES), a symmetric encryption calculation took on by the U.S. government, is broadly utilized for getting touchy information, like monetary exchanges and government correspondences.

Quantum cryptography, in view of the standards of quantum mechanics, expects to utilize quantum peculiarities to make secure correspondence channels resistant to listening in.

Cryptanalysis is the investigation of cryptographic frameworks determined to break them or tracking down shortcomings. It includes different techniques, including savage power assaults, recurrence investigation, and taking advantage of numerical weaknesses.

Cryptography assumes a vital part in different fields past correspondence, including digital money, getting clinical records, safeguarding protected innovation, and guaranteeing information security in distributed computing.

COOKING

Cooking is quite possibly of the most seasoned human action. Proof recommends that early people started cooking over controlled fires around a long time back.

Different cooking strategies modify the wholesome substance of food. For example, bubbling can filter supplements into the water, while barbecuing or cooking can hold more supplements.

Maillard response, a synthetic response between amino acids and sugars at high temperatures, is liable for the carmelizing of food, making complex flavors and smells.

The presentation of new flavors and fixings during the Time of Investigation changed worldwide cooking styles. Food sources like tomatoes, potatoes, and stew peppers began from the Americas and spread around the world.

Cooking is a multisensory experience. Smell, sight, taste, contact, and, surprisingly, sound (like sizzling) assume parts by they way we see and appreciate food.

Sub-atomic gastronomy is a part of cooking that investigates the physical and compound changes of fixings. It frequently includes imaginative procedures and devices to make special surfaces and flavors.

The idea of umami, an exquisite taste frequently depicted as the fifth taste close by sweet, harsh, pungent, and unpleasant, assumes a critical part in upgrading flavors in dishes.

Maturation, a conventional food protection procedure, broadens the timeframe of realistic usability of food as well as adds complex flavors and valuable probiotics to specific food varieties like kimchi, yogurt, and sauerkraut.

Cooking can be restorative and reflective, permitting individuals to communicate inventiveness and interface with others through shared dinners.

The way of life and history of a district fundamentally impact its food, prompting different cooking styles, fixings, and customary dishes around the world.

PROGRAMMING

The main software engineer was Ada Lovelace, who composed a calculation for Charles Babbage's Logical Motor during the 1800s, making her the world's most memorable developer.

The expression "bug" began from a real moth found inside the Harvard Imprint II PC in 1947. This occurrence prompted the utilization of the expression "troubleshooting" for fixing blunders in code.

The primary undeniable level programming language was Fortran (Recipe Interpreter), created during the 1950s for logical and designing computations.

JavaScript, regardless of its name, has no connection to Java. It was created by Brendan Eich in only ten days and was at first called Mocha, later renamed to LiveScript, lastly JavaScript for promoting reasons.

Python, a famous programming language known for its lucidness and flexibility, was named after the English parody show "Monty Python's Flying Carnival," not the reptile.

The Y2K bug was a programming issue in the last part of the 1990s where many expected that PC frameworks would fall flat while progressing from 1999 to 2000 because of the two-digit year design. Broad endeavors were made to forestall possible disturbances.

The primary PC infection, called the Creeper, was created in the mid 1970s. It didn't cause harm however showed a message on contaminated PCs saying, "I'm the creeper, get me if possible!"

The idea of Article Arranged Programming (OOP) was spearheaded by Ole-Johan Dahl and Kristen Nygaard during the 1960s with the programming language Simula.

Git, a variant control framework generally utilized by software engineers, was made by Linus Torvalds in 2005 for dealing with the Linux part's turn of events.

The principal compiler, a program that deciphers code written in a significant level language into machine code, was created by Beauty Container during the 1950s for the A-0 Framework programming language.

LOGO DESIGN

Logos have been around for a really long time, tracing all the way back to old civic establishments that pre-owned images to address their personalities or items.

Logos mean to convey a brand's personality briefly and importantly, frequently through images, typography, or a mix of both.

Variety brain science assumes a critical part in logo plan. Various tones inspire explicit feelings and discernments, impacting how individuals see a brand.

Logos develop over the long haul. Numerous famous logos have gone through different updates to remain applicable and adjust to changing patterns or brand methods of reasoning.

Effortlessness is key in viable logo plan. Essential logos are frequently simple and effectively unmistakable at different sizes and on various mediums.

Logos can rise above language hindrances. A very much planned logo can impart a brand's message across different societies without the requirement for words.

Responsive plan contemplations are pivotal in present day logo creation. Logos need to adjust consistently across advanced stages, print media, stock, and different screen sizes.

The most common way of planning a logo includes broad exploration, conceptualizing, outlining, criticism, and refinement prior to showing up at the last plan that embodies a brand's pith.

Inventions

Creatures arrive in a mind blowing scope of sizes, from the little honey bee hummingbird (2.2 inches/5.5 cm long) to the blue whale, the biggest creature on The planet (up to 100 feet/30 meters in length).

The copy octopus can mimic the appearance and conduct of different creatures, like lionfish, flatfish, and ocean snakes, as a guard system.

The tardigrade, or water bear, is an amazingly tough infinitesimal creature that can endure outrageous circumstances, including outrageous temperatures, radiation, and, surprisingly, the vacuum of room.

The tongue of a blue whale can weigh however much that an elephant, and its heart can be the size of a little vehicle, making it perhaps of the most monstrous animal on earth.

The gun shrimp has a specific paw that snaps shut so rapidly it makes a cavitation bubble, prompting a noisy commotion and temperatures nearly as warm as the outer layer of the sun.

The bombardier scarab can deliver a burning compound shower as a guard component. When undermined, it blends synthetic compounds in its mid-region, causing a hazardous response.

The archerfish can destroy bugs by spitting a stream of water with exceptional exactness, thumping prey into the water to eat.

The axolotl, a kind of lizard, has the inconceivable capacity to recover its appendages, spinal string, heart, and other body parts, making it a priceless resource for logical exploration.

A few types of subterranean insects go about as ranchers, developing parasite gardens. They gather leaves and use them to grow a parasite that fills in as their essential food source.

The wheel, considered perhaps of humankind's most urgent development, traces all the way back to around 3500 BC and altered transportation and hardware.

The print machine, imagined by Johannes Gutenberg in the fifteenth hundred years, empowered large scale manufacturing of books and altogether affected the spread of information and data.

The light, frequently connected with Thomas Edison, was really a cooperative development. Edison's group fostered the principal monetarily pragmatic glowing light.

A few innovations have steered history, for example, the steam motor, which powered the Modern Upset and changed assembling and transportation.

HEALTHY FOOD

Supplement thick food sources like organic products, vegetables, lean proteins, and entire grains give fundamental nutrients, minerals, and cancer prevention agents significant for generally speaking wellbeing.

Blueberries are known as a superfood because of their elevated degrees of cell reinforcements, which assist with combatting oxidative pressure and lessen the gamble of constant illnesses.

Nuts, like almonds and pecans, are plentiful in sound fats, fiber, and different nutrients, advancing heart wellbeing and lessening the gamble of coronary illness.

Greasy fish like salmon, mackerel, and sardines are magnificent wellsprings of omega-3 unsaturated fats, essential for cerebrum wellbeing, diminishing aggravation, and supporting heart wellbeing.

Quinoa is a without gluten grain plentiful in protein, fiber, and different nutrients and minerals, making it a nutritious option in contrast to customary grains.

Avocados are stacked with heart-solid monounsaturated fats, fiber, potassium, and different nutrients, adding to further developed cholesterol levels and heart wellbeing.

Mixed greens like spinach, kale, and Swiss chard are loaded with nutrients, minerals, and cell reinforcements, supporting bone wellbeing and diminishing the gamble of ongoing infections.

Vegetables, including beans, lentils, and chickpeas, are wealthy in protein, fiber, and fundamental supplements, helping with weight the executives and advancing stomach related wellbeing.

Entire grains like oats, earthy colored rice, and quinoa contain fiber, nutrients, and minerals, adding to better assimilation, diminished irritation, and further developed heart wellbeing.

SEA FOOD

An octopuses have an extraordinary capacity to change their skin tone and surface to match their environmental elements, helping them cover and stow away from hunters or prey.

The copy octopus has the wonderful expertise to impersonate other ocean animals, taking on their appearance and developments to sidestep hunters.

A few types of jellyfish are naturally interminable, meaning they can hypothetically live endlessly by returning to a prior progressive phase when harmed or maturing.

Dolphins are exceptionally shrewd and show mindfulness. They can perceive themselves in mirrors, a quality common with a couple of different creatures like gorillas and elephants.

The gun shrimp makes a cavitation bubble when it snaps its particular paw shut, creating a sound arriving at up to 218 decibels — stronger than a discharge — and staggering its prey.

The blue-ringed octopus, notwithstanding its little size, conveys a toxin deadly to people. It's unimaginably venomous, containing poisons that can cause loss of motion and even passing.

Ocean otters are one of only a handful of exceptional creatures known to utilize devices. They use rocks to air out shellfish and store their number one rocks in pockets of skin under their lower arms for supervision.

THE END

CONGRATULATIONS